ANALYTICAL
METHOD
DEVELOPMENT
AND
VALIDATION

ANALYTICAL METHOD DEVELOPMENT AND VALIDATION

Michael Swartz

Waters Corporation
Milford, Massachusetts

Ira S. Krull

Northeastern University
Boston, Massachusetts

Waters

MARCEL DEKKER, INC.   NEW YORK · BASEL · HONG KONG

ISBN: 0-8247-0115-1

The publisher offers discounts on this book when ordered in bulk quantities. For more information, write to Special Sales/Professional Marketing at the address below.

This book is printed on acid-free paper.

MARCEL DEKKER, INC.
270 Madison Avenue, New York, New York 10016
http://www.dekker.com

Current printing (last digit):
10 9 8 7 6 5 4 3 2 1

PRINTED IN THE UNITED STATES OF AMERICA

Preface

This book was designed to present and discuss a rationale for the process of successful development of HPLC-based analytical methods, their optimization, and eventual validation. Although the *U.S. Pharmacopeia* (USP) addresses the topic of method validation and although various other sources cover HPLC method development and optimization on a variety of levels, the literature is ambiguous regarding the overall formal process that combines method development and validation. There are many textbooks on analytical chemistry and instrumental methods of analysis, but they do not address the specifics of this topic. Also, although there are many citations on method validation for HPLC in the literature, very few of these papers discuss approaches to method development and optimization that incorporate method validation. Therefore, in spite of the fact that many analytical chemists in academia, industry, and government laboratories spend a good deal of their time attempting to develop new or improved validated methods for specific analytes, there is little in the literature to guide them along these pathways. In addition, as a result of the International Conference on Harmonization, (ICH), new guidelines are in preparation that, it can be anticipated, will eventually be incorporated into the USP. We attempt here to define and delineate the individual steps of method development, optimization, and validation, and to show how these steps can be integrated into a formal process, reflecting both the current USP regulations and ICH contributions.

In *Analytical Method Development and Validation*, the subject of developing and optimizing an HPLC method is presented, culminating in a step-by-step guideline. Next, the process of validation is discussed, starting with instrument qualification and concluding with system suitability. Validation is presented on the basis of not only the currently accepted USP terminology and methodology but also according to ICH guidelines. Then, an analytical method validation protocol is proposed. Finally, a bibliography points the reader to additional literature of interest. We hope that this book will provide analytical chemists with direction and guidance to simplify the overall process of method development, optimization, and validation.

Michael E. Swartz
Ira S. Krull

Contents

Tables and Figures

11

Abbreviations

Abbreviation	Definition
AOAC	Association of Official Analytical Chemists
ASTM	American Society for Testing and Materials
CAS	Chemical Abstracts Service
CE or HPCE	High-performance capillary electrophoresis
DQ	Design or development qualification
EPA	Environmental Protection Agency
FDA	Food and Drug Administration

Abbreviation	Definition
FL	Fluorescence
GC	Gas chromatography
GLP	Good laboratory practices
cGMP	Current good manufacturing practices
HPLC	High-performance liquid chromatography
ICH	International Conference on Harmonization
ICP	Inductively coupled plasma
IQ	Installation qualification
LCED	Liquid chromatography-electrochemical detection
LCL	Lower control limit
LOD	Limit of detecttion
LOQ	Limit of quantitation
MQ	Maintenance qualification
MS	Mass spectrometry
NF	National Formulary
NMR	Nuclear magnetic resonance
NIST	National Institute of Standards and Technology
OQ	Operational qualification
PQ	Performance qualification
SOPs	Standard operating procedures
SRM	Standard reference material
UCL	Upper control limit
USP	*United States Pharmacopeia*
UV	Ultraviolet
VIS	Visible

ANALYTICAL METHOD DEVELOPMENT AND VALIDATION

Introduction

Analytical chemistry, which is both a theoretical and a practical science, is practiced in a large number of laboratories in many diverse ways. Methods of analysis are routinely developed, improved, validated, collaboratively studied, and applied. Compilations of methods appear in large compendia, such as the *U.S. Pharmacopeia* (USP), *EPA Handbook of Methods for Environmental Pollutants*, *Official Methods of Analysis* [Association of Official Analytical

Chemists (AOAC)], and others. Instrumental approaches are developed by industrial and clinical/environmental laboratories to meet regulatory requirements, as well as by regulatory agencies and other federal, state, and local agencies and laboratories. It can be reasonably asserted that chemists practice analytical chemistry, in one form or another, more than any other branch of chemistry. Surely this is true in regulated industries, such as the pharmaceutical industry, and by agencies that regulate them, such as the U.S. Food and Drug Administration (FDA).

Given the widespread use of analytical methods, it is noteworthy that apparently there are no formally accepted guidelines or formats for the overall process of design, development, optimization, and validation of analytical methods. Books written on analytical chemistry do not always fully and specifically describe the overall process. Academic analytical chemists, more than their industrial colleagues, seem to avoid in-depth discussions regarding the development of new methods; instead, they follow development procedures of variable rigor and imply that these are correct and valid. In addition, many academicians develop and optimize analytical methods using pure standards, leaving validation or application to those in the field. This situation represents an unfortunate evolution of the science, because too many analytical methods lacking full and adequate validation have been introduced into the open literature. These methods are allowed to assume the aura of validity and authenticity, when, in practice, they may never really be useful with samples that are typically encountered.

This book presents the basic considerations needed to arrive at an optimized, valid, and reliable analytical method, as well as guidelines for the evaluation of the final reliability of such methods. These considerations include conception and design of the new method: development of the initial method, optimization and improvement of this method, characterization of the optimized method, and method validation.

In a strict sense, however, method validation is a subset of the total validation process, which encompasses many different aspects. Therefore, this process relative to method validation must be addressed first. In addition, while this book is written primarily from an FDA focus, reflecting current USP guidelines, any treatment of the topic of validation would be remiss if it did not address recent developments in the field as a result of the International Conference on Harmonization of Technical Requirements for Registration of Pharmaceuticals for Human Use (ICH). It is assumed, however, that the reader has some familiarity with FDA, USP, and ICH guidelines.

1.1. THE VALIDATION PROCESS

Validation is a process that consists of at least four distinct steps: (1) software validation, (2) hardware (instrumentation) validation/qualification, (3) method validation, and (4) system suitability, as represented in Figure 1. The process begins with validated software and a qualified system. Then a validated method using the qualified system is developed. Finally, total validation is achieved by defining system suitability. Each step is critical to the overall success of the process.

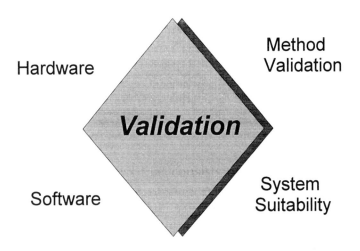

FIGURE 1   The validation process.

Prior to undertaking the task of method validation, it is necessary to invest some time and energy up front to ensure that the analytical system itself is validated, or *qualified*. Qualification is a subset of the validation process that verifies module and system performance prior to the instrument being placed on-line. If the instrument is not qualified prior to use and a problem is encountered, the source of the problem will be difficult to identify. Figure 2 depicts a timeline of the events that make up the validation process. The steps are discussed in detail in the following sections.

1.2. QUALIFICATION

As illustrated in Figure 2, validation begins at the vendor's site, as part of a structural validation stage. During this stage, the instrument and software are developed, designed, and produced in a validated environment according to good laboratory practices (GLP), current good manufacturing practices (cGMP), and/or ISO 9000 standards (TickIT). During the functional validation or qualification stage, the installation qualification (IQ), operational qualification (OQ), and performance qualification (PQ) are performed (as outlined in Sections 1.2.1–1.2.3). After the instrument is placed on-line in the laboratory, and after a set period of use, regulations require maintenance followed by calibration and standardization, sometimes referred to as maintenance qualification (MQ) or maintenance procedures. Each laboratory should have standard operating procedures (SOPs) in place that define the period of use (usually defined as a reasonable interval during which the instrument operates without any loss in functional performance) and the procedures for placing the instrument on-line following maintenance. Logs should be used to keep track of instrument maintenance and to maintain spare parts inventories.

1.2.1. Installation Qualification (IQ)

The IQ process can be divided into two steps, *preinstallation* and *physical installation*. During preinstallation, all the information pertinent to the proper installation, op-

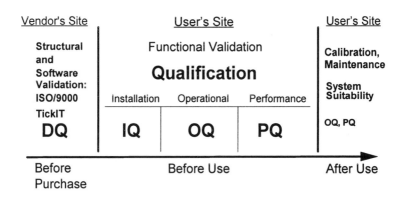

FIGURE 2   The validation timeline.

eration, and maintenance of the instrument is reviewed. Site requirements and the receipt of all of the parts, pieces, manuals, etc., necessary to perform the installation are confirmed. During physical installation, serial numbers are recorded, and all of the fluidic, electrical, and communication connections are made for components in the system. Documentation describing how the instrument was installed, who performed the installation, and other miscellaneous details should be archived.

1.2.2. Operational Qualification (OQ)

The OQ process ensures that the specific modules of the system are operating according to the defined specifications for accuracy, linearity and precision. This process may be as simple as verifying the module's self-diagnostic routines, or it may be performed in more depth by running specific tests, for example, to verify detector wavelength accuracy, flow rate, or injector precision.

1.2.3. Performance Qualification (PQ)

The PQ process verifies system performance. PQ testing is conducted under actual running conditions across the anticipated working range. In practice, however, OQ and PQ are frequently performed together, particularly for linearity and precision (repeatability) tests, which can be conducted more easily at the system level. For HPLC, the PQ test should use a method with a well-characterized analyte mixture, column, and mobile phase. It should incorporate the essence of the System Suitability section of the general chromatography chapter (<621>) from the USP. Again, proper documentation supporting the PQ process should be archived.

Method Development, Optimization, and Validation Approaches

2.1 METHOD DEVELOPMENT

Today, the development of a method of analysis is usually based on prior art or existing literature, using the same or quite similar instrumentation. It is rare today that an HPLC-based method is developed that does not in some way relate or compare to existing, literature-based approaches. The development of any new or improved method usually tailors existing approaches and instrumentation to the current analyte, as well as to the final needs or requirements of the method. Method development usually requires selecting the method requirements and deciding on what type of instrumentation to utilize and why. In the development stage, decisions regarding choice of column, mobile phase, detector(s), and method of quantitation must be addressed. In this way, development considers all the parameters pertaining to any method.

There are several valid reasons for developing new methods of analysis:

• There may not be a suitable method for a particular analyte in the specific sample matrix.

• Existing methods may be too error-, artifact-, and/or contamination-prone, or they may be unreliable (have poor accuracy or precision).

• Existing methods may be too expensive, time consuming, or energy intensive, or they may not be easily automated.

• Existing methods may not provide adequate sensitivity or analyte selectivity in samples of interest.

• Newer instrumentation and techniques may have evolved that provide opportunities for improved methods, including improved analyte identification or detection limits, greater accuracy or precision, or better return on investment.

- There may be a need for an alternative method to confirm, for legal or scientific reasons, analytical data originally obtained by existing methods.

Next, it is necessary to translate the goals of the method into a method development design. Goals for a new or improved analytical method might include the following:

- Qualitative identification of the specific analyte(s) of interest, providing some structural information to confirm "general behavior" (e.g., retention time, color change, pH).
- Quantitative determination, at trace levels when necessary, that is accurate, precise, and reproducible in any laboratory setting when performed according to established procedures.
- Ease of use, ability to be automated, high sample throughput, and rapid sample turnaround time.
- Decreased cost per analysis from using simple quality assurance and quality control procedures.
- Sample preparation that minimizes time, effort, materials, and volume of sample consumed.
- Direct output of qualitative or quantitative data to laboratory computers in a format usable for evaluation, interpretation, printing out, and transmission to other locations via a network.

Once the instrumentation has been selected, based on the criteria suggested above, it is important to determine "analyte parameters" of interest. To develop a method, it is necessary to consider the properties of the analyte(s) of interest that may be used to advantage and to establish optimal ranges of analyte parameter values. Examples include chromatographic capacity factors (a target range of $k' = 2$ to 10 is generally desirable), UV-VIS/FL wavelengths of detection, mass/charge ratios to be scanned, and optimal emission wavelengths. Such information may already be available in the literature for the analyte or related compounds. If not, it may be necessary to ascertain analyte separation–detection parameters using standards and suitable analytical methods. If this is the case, the purity of the analytical standard must be firmly demonstrated.

Once the instrumentation has been assembled and analyte parameters have been considered, standards should be used for the continued development, optimization, and preliminary evaluation of the method. Initial analytical figures of merit should be ascertained, including sensitivity, measured as response per amount (concentration or mass) injected; limits of detection; limits of quantitation; linearity of calibration plots; and multi-detector ratios [e.g., dual-wavelength UV-VIS ratios or dual-electrode ratios in liquid chromatography-electrochemical detection (LCEC)]. It is important that method development be performed using only analytical standards that have been well identified and character-ized, and whose purity is already known. Such precau-tions will prevent problems in the future and will re-move variables when one is trying to optimize or im-prove initial conditions during method development.

2.2 OPTIMIZATION

During the optimization stage, the initial sets of condi-tions that have evolved from the first stages of develop-ment are improved or maximized in terms of resolution, peak shape, plate counts, asymmetry, capacity, elution time, detection limits, limit of quantitation, and overall ability to quantify the specific analyte of interest. When optimizing any method, an attempt should be made to provide analytical *figures of merit* which are needed to meet the assay requirements defined at the initial stages of method development. In other words, the required detection limits, limits of quantitation, accuracy and pre-cision of quantitation, and specificity must be defined. Without adequate and definitive requirements, it is diffi-cult to optimize any analytical method.

Results obtained during optimization must be evaluated against the goals of the analysis set forth by the analytical figures of merit. This evaluation may re-veal that additional development and optimization are needed to meet some of the initial method requirements.

Once some initial analytical figures of merit are available, it is important to pause and consider the outcome. Does the new method offer advantages over existing methodology in terms of the goals of the analysis? How does the new method compare with currently accepted methods of analysis for the analyte of interest? If the method seems to offer significant advantages, then its development warrants additional time and effort. If no advantages are apparent, it may be necessary to re-evaluate the entire approach to decide whether additional time and effort are justified.

At this point, it is also possible to evaluate the scope of the new method; that is, will it be applicable only to the main analyte of interest, or also to related analytes or entire classes of such compounds? Obviously, a method that is able to determine only one specific analyte has far less utility than alternative methods that can analyze that analyte selectively and many others under slightly different working parameters.

If the initial analytical data derived from the new method appears promising, it is crucial to evaluate its performance quantitatively. Initially, most work on method development and optimization is performed with analytical standards. In general, the analytical figures of merit generated to evaluate the method are also derived using standards. The scope of the method evaluation should be broad enough to include generation of information that is immediately usable for confirmation of the analyte in any sample, for example UV absorbance and emission spectra, electrochemical response behavior (e.g., voltammograms), or mass spectra. Optimization of the method should yield maximized sensitivity, peak symmetry, minimized detection and quantitation limits, a wide linear dynamic range, and a high degree of accuracy and precision. Other potential optimization goals include baseline resolution of the analyte of interest from other sample components, unique peak identification, on-line demonstration of purity, and interfacing of computerized data for routine sample analysis. Absolute quantitation should use simplified methods that require minimal sample handling and analysis time. Optimization criteria must be determined with cognizance of the goals common to any new method, such as reduced analysis time and cost, and accurate identification of the analyte.

Optimization of the method can follow either of two general approaches: (1) manual or (2) computer driven. The manual approach involves varying one experimental variable at a time, while holding all others constant, and recording changes in response. The variables might include flow rates, mobile or stationary phase composition, temperature, detection wavelength, and pH. This univariate approach to system optimization is slow, time consuming, and potentially expensive. However, it may provide a much better understanding of the principles and theory involved, and of the interactions of the variables. In the second approach, computer-driven automated methods development, efficiency is optimized while experimental input is minimized. Computer-driven automated approaches can be applied to many applications. In addition, they are capable of significantly reducing the time, energy, and cost of virtually all instrumental methods development.

To obtain analytical figures of merit, the method should be optimized with respect to selectivity, accuracy, precision, limit of detection, limit of quantitation, linearity, and range. There are other important optimization parameters that need to be met (such as robustness and ruggedness) but whose measurement and evaluation are better addressed by method validation than by optimization.

How does the analyst really know that the new method has been adequately optimized and is ready for actual sample applications? To verify that the optimized method satisfies the goals of unequivocal analyte identification and quantitation, improved quantitative accuracy and precision, faster sample turn around time, absence of interference, and automation, certain general criteria can be considered:

- Chromatographic resolution is adequate.
- For most samples, limits of detection are lower by at least one order of magnitude than needed.
- Calibration plots are linear over several orders of magnitude, beginning with limits of quantitation.

- Sample throughput is increased, with minimal instrument equilibration (before run and from run to run).
- Sample preparation before analysis is minimized.
- Interference is minimized and identified, and approaches are established to circumvent such problems.
- Data is acquired by computer or reporting integrator and can be manipulated, translated, interpreted, printed, or stored in various forms. If applicable, computer software allows for rapid acquisition, storage, and manipulation of data in either a standalone workstation or client–server environment.
- Reproducibility of analytical figures of merit is demonstrated, with acceptable accuracy and precision.
- Cost per analysis is minimized.

System optimization is one of the most time- and energy-consuming parts of the overall methods development procedure. It requires an iterative procedure, constant replication, and the acquisition of a large amount of quantitative data. Too often, optimization results in a method that meets the immediate requirements of the analyst but ignores possible future needs. Ideally, the analyst should optimize each new method to the fullest practical extent, in order to ensure a broad utility of the method and obviate the repetition of experiments for future method development.

2.3 METHOD VALIDATION APPROACHES

It is difficult to completely separate method development and optimization from validation; these areas often overlap. For the purpose of this discussion, however, it is possible to separate optimization from validation. In the validation stage, an attempt should be made to demonstrate that the method works with samples of the given analyte, at the expected concentration in the matrix, with a high degree of accuracy and precision. Other validation criteria exist, as defined and elaborated upon later, but complete method validation can occur only after the method is developed and optimized. In validation studies, suitability of the final method for the given analyte and a select sample matrix is demonstrated, using specified instrumentation, samples, and data handling; ultimately, the method can be transferred from one laboratory to another that is suitably equipped and staffed. A method that provides all or most of the original method requirements is deemed optimized and becomes ready for validation.

A great deal has been written on general method validation guidelines, and there are numerous approaches that can be employed. There is no single validation approach that must always be employed for a new method; the analyst's primary concern should be to select an approach that will prove to be a true validation. Acceptance of any new method by others in the field will depend on the specific validation approaches used. It is the responsibility of the individual analyst to select the correct validation method(s). Validation approaches include the zero-, single-, and double-blind spiking methods; interlaboratory collaborative studies; and comparison with a currently accepted (compendium) method.

2.3.1 The Zero-Blind Method

The zero-blind approach involves a single analyst using the method with samples at known levels of analyte to demonstrate recovery, accuracy, and precision. The method is subject to analyst bias, and though the method is, in general, fast, simple, and useful, it leads to subjective results and doubt on the part of the unbiased reviewer or end user. However, as a first approximation and a demonstration of validation potential requiring minimal time, manpower, samples, and cost, a zero-blind study is a good place to start the overall validation process. Clearly, if this approach fails to validate a method, then there is no reason to proceed with further validation of the method.

2.3.2 The Single-Blind Method

The single-blind approach involves one analyst preparing samples at varying levels unknown to a second analyst, who also analyzes the samples. The results are then compiled and compared by the first analyst. Although this approach is unbiased at the start, it loses its blindness at the most crucial stage—when both sets of data are compared. While perhaps more valuable and believable than the zero-blind approach, the single-blind approach still invites bias on the part of the first analyst to bring two sets of data into better agreement. This approach is appropriate at the very start of the method validation, after the single-blind approach has proven successful, but before one decides to involve additional analysts or management.

2.3.3 The Double-Blind Method

The double-blind approach involves three analysts. The first analyst prepares samples at known levels, the second does the actual analysis, and the third analyst (or administrator) compares both sets of data received separately from the first two analysts. Neither the first or second analyst has access to the set of data generated by the other. This double-blind approach is the most objective approach, assuming no bias on the part of the third analyst.

2.3.4 The Analysis of Standard Reference Materials

The analysis of a standard reference material (SRM) or an authenticated sample is a generally accepted method of validation. The *USP*, NIST, and other, private organizations specialize in preparing, guaranteeing, and marketing standard reference materials of various analyte species in different sample matrices. It may be necessary, however, to contract the preparation of a unique sample in a particular matrix in order to utilize this procedure for method validation. When using SRMs, the analyst must demonstrate that the method provides accurate and precise measurements of the analyte in a particular sample matrix. Analyst bias can also be an issue, especially when the analyst knows the amounts and levels of the SRM.

2.3.5 The Interlaboratory Collaborative Study

The interlaboratory collaborative study is perhaps the most widely accepted procedure to validate any new analytical method, but it suffers from serious practical drawbacks. The collaborative approach is costly and time consuming; it can take years from start to finish. During that time, the analysts may have to expend considerable effort coordinating the process, shipping samples and receiving results, statistically analyzing and interpreting the results, and then finally interpreting and verifying the data. Although the approach is operator-dependent (generating laboratory-to-laboratory variability), when all laboratories involved come up with overlapping quantitative values in comparison with known levels present, the method is generally accepted as full validation. This approach is rarely employed when a method is being described for the first time in the literature.

2.3.6 Comparison with a Currently Accepted Method

Comparison with a currently accepted analytical method is yet another validation approach. This is usually done by a single analyst, but it can be done by two analysts using a split sample. This approach uses results from the currently accepted method as verification of the new method's results. Agreement between results initially suggests validation. Disagreement is a serious cause for concern of future acceptability of the new method. However, disagreement could also suggest that the currently accepted method is invalid, creating additional problems. If the analyst can prove that the currently accepted method is indeed invalid, the analyst must then initiate an alternative approach to validate the new method.

The question will eventually arise as to how many samples should be analyzed in any validation approach—1, 2, 6, 12? In general, the more the better, and the greater the variety of samples and variation in the concentration range the better. Ideally, the method should be validated for the analyte using several different sample types, with several of each type determined separately for statistical and validation purposes. A single, zero-blind or a single, single-blind study is obviously less meaningful and less acceptable than an interlaboratory collaborative, true double-blind study of several sample matrices at widely different concentration levels. Initial validation approaches are generally less rigorous and demanding than ones performed for standard reference material (SRM) development.

2.4. STEP-BY-STEP HPLC METHOD DEVELOPMENT, OPTIMIZATION, AND VALIDATION: AN OUTLINE

This section provides, in outline form, a "checklist" for method development, optimization, and validation approaches. It should be considered in combination with other sections (2.3, 3.1, and 3.2) to complete the overall validation process. Although the outline is aimed primarily aimed at chromatographic methods of analysis, the overall approach should be amenable and adaptable to other forms of testing, such as physical testing, performance testing, and other chemical testing.

Because documentation should start at the very beginning of the method development process, a system for full documentation of the development studies must be established. All data relating to these studies must be recorded in laboratory notebooks or an electronic data base.

1. Analyte Standard Characterization

  a. Assemble all known information about the analyte and its structure, physical and chemical properties, toxicity, purity, hygroscopicity, solubility, and stability.

b. Determine if a standard analyte (100% pure) is available, and if so, the quantity available, the storage equipment required (refrigerator, dessicator, freezer, aqueous or organic solutions, etc.), and the disposal requirements.

c. If there are multiple components to be analyzed in the sample matrix, note the number of components; assemble data and determine the availability of standards for each one.

d. Take particular note when samples are limited (small volume or mass) or an analyte is present at trace levels.

e. Consider the availability of standards for degradation products, possible impurities, and synthetic precursors. The purity of all standards to be used in method development, optimization, and validation should be demonstrated and documented. Reference standards (USP, NIST, EPA, etc.) should be used whenever possible.

f. Consider only those methods (MS, GC, CE, HPLC, etc.) that are compatible with sample stability.

2. Method Requirements

a. Consider the goals or requirements of the analytical method that need to be developed, and define the analytical figures of merit. Define the required detection limits, selectivity, linearity, range, and accuracy and precision.

b. Consider the eventual use of the method, its adaptability to other uses, and any regulatory requirements that need to be met. Additional requirements might include sample throughput (time, effort, money, and labor), analysis time, available instrumentation (isocratic vs. gradient-elution HPLC), instrument limitations (pressure and solvents), and cost per analysis.

3. *Literature Search and Prior Methodology*

a. Search the literature for all types of information related to the analyte. Determine if its synthesis, physical and chemical properties, solubility, or relevant analytical methods have been described. Consult books, periodicals, chemical manufacturers, and regulatory agency compendia, such as USP/NF, AOAC, and ASTM publications. Use Chemical Abstracts Service (CAS) automated/computerized literature searches, if possible.

b. Determine if any analytical work on the analyte has ever been done within the company, and if so, compile data, results, reports, memos, and publications. Communicate with colleagues who might know more about prior analytical work performed in-house that was published (or not), reported, or abstracted.

4. *Choosing a Method*

a. Adaptation is more efficient than "reinventing the wheel." Determine if any of the reported methods from the literature are adaptable to the current laboratory setting and future needs.

b. Using the information in the literature and prior methodology, adapt and modify methods. Determine if it is necessary to acquire additional instrumentation to reproduce, modify, improve, or validate existing methods for in-house analytes and samples.

c. Whenever possible, adopt sample preparation and instrument conditions (e.g., HPLC or HPCE) to take advantage of the latest methods and instrumentation. Consider automation of instruments with computer interfacing.

d. If there are no prior methods for the analyte in the literature, work from analogy to investigate compounds that are similar in structure and properties. There is usually one compound for which analytical methods already exist that is similar to the analyte of interest.

5. *Instrument Setup and Initial Studies*

a. Set up the required instrumentation. Verify installation and operational and performance qualifications of instrumentation using laboratory standard operating procedures (SOPs).

b. Instrumentation should be interfaced to a computer and should be microprocessor controlled. Make use of method optimization software if it exists for the method, and utilize computer software for data acquisition and analysis.

c. Always use new consumables (e.g., solvents, filters, and gases). For example, never start method development on an HPLC column that has been used before.

d. Prepare the analyte standard in a suitable injection/ introduction solution and in known concentrations and solvents. It is important to start with an authentic, known standard rather than with a complex sample matrix. If the sample is extremely close to the standard (e.g., bulk drug), then it is possible to start work with the actual sample.

e. Begin the analysis using the analytical conditions described in the existing literature.

f. Evaluate feasibility of method with regard to the analytical figures of merit obtained.

6. *Optimization*

a. If initial analytical results are less than ideal, begin the optimization process, keeping in mind the goals of the method. Utilize computer-based optimization software whenever possible.

b. During optimization, pay particular attention to experimental design. Consult the literature on this subject, if possible. Change one parameter at a time, and isolate sets of conditions, rather than use a trial-and-error approach. Work from an organized, methodical plan, and document every step (keep a lab notebook) in case of dead ends.

c. Keep in mind sample introduction requirements considered unique to a particular instrument or method of analysis. Consider what form the sample should take for final analysis, and maintain proper sample preparation and final injection/introduction matrix compatibility.

7. *Demonstration of Analytical Figures of Merit with Standards*

a. Document the originally determined analytical figures of merit [limit of quantitation (LOQ), limit of detection (LOD), linearity, time per analysis, cost, sample preparation, etc.). Up to this point, all work has been performed with standards (if available). Demonstrate optimized analytical figures of merit for the standard first, before proceeding to actual sample work. If the required analytical figures of merit cannot be met by the standard, sample analysis is pointless. When the analytical figures of merit are optimized and documented, including a standardization of such things as integration parameters and any statistical treatment of the data (if necessary), then analysis of samples can begin.

b. At the onset of sample analysis, document peak homogeneity by performing appropriate peak identification studies (e.g., PDA, MS, FTIR). Include blank and control samples along with a defined system suitability standard. Ensure the absence of false positives or false negatives arising from artifacts in the method, the chemicals/solvents used, or instrumental aberrations.

8. *Evaluation of Methods Development with Actual Samples: Derivation of Figures of Merit*

a. Working with actual samples, perform the sample preparation steps to ensure detectability of the analyte peak apart from all other potential interferences and contaminants.

b. Optimize selective sample preparation to achieve a purer final injection solution, resulting in better accuracy, precision, reproducibility, and final validation with real samples. The goal is a simple preparation step. Ideally, a "dilute and shoot" sample preparation will minimize time and cost of the overall analysis. At the same time, one must remember to provide an injection solution that is compatible with the HPLC/HPCE/ICP/MS system.

c. ensure that the sample solution leads to unequivocal, absolute identification of the analyte peak of interest apart from all other matrix components.

9. *Validation of Figures of Merit*

a. Validate the method once it has been developed and optimized. Regulatory laboratories (such as the FDA) perform method validation by evaluating and documenting the USP "eight steps of method validation."

These include precision, accuracy, linearity range, LOD, LOQ, specificity, ruggedness, and robustness (Chapter 3). Some of these steps (linearity range, LOD, LOQ, specificity) have already been evaluated as analytical figures of merit. However, robustness must also be measured.

10. *Determination of Percent Recovery of Actual Sample and Demonstration of Quantitative Sample Analysis*

a. Determine percent recovery of spiked, authentic standard analyte into a sample matrix that is shown to contain no analyte.
b. Show reproducibility of recovery (average ± standard deviation) from sample to sample and whether recovery has been optimized. It is not necessary to obtain 100% recovery as long as the results are reproducible and known with a high degree of certainty.
c. Develop a method of quantitation that will take percent recovery into account. Possible methods include standard additions, external/internal standard, and isotopic dilution. Indicate or determine which of these methods of quantitation provides the most accurate and precise quantitative data.

11. *Method Validation*

a. Perform zero-blind studies to demonstrate that known levels can be accurately and precisely determined in a real sample (quantitation).

b. Perform double-blind studies to further demonstrate the quantitative accuracy and precision of the overall method.

c. Demonstrate repeatability of analytical results, within a single laboratory, of retention time, peak shape/asymmetry, peak height, plate counts, efficiency, reduced plate height, and so forth. Also, demonstrate analytical figures of merit, including robustness, for real samples, again as for an authenticated standard alone.

d. Demonstrate reproducibility (ruggedness), from lab to lab, analyst to analyst, instrument to instrument, and so on, as required.

e. Perform additional validation using an authentic SRM of the analyte in the sample matrix of major interest. If no SRM is available, have one synthesized by an outside, contract laboratory that will guarantee authenticity of composition and identity. When obtaining standards by custom synthesis, confirm their purity and identity by LC-MS, NMR, and so forth. Synthesize your own SRMs by spiking samples shown to have no background levels of analyte at known, varying levels of 100% standard analyte.

f. Other methods of validation are available. Undertake one or more of these more elaborate approaches if double-blind approaches and SRMs provide accurate and precise quantitation results.

12. *Preparation of Written Protocols and Procedures (Method Manual)*

   a. Prepare written protocols and procedures for other laboratories to follow based on all of the above method development and optimization studies. Written procedures satisfy regulatory agencies and facilitate method transfer. Include all possible specifics of the chemicals, reagents, instrumentation (with operational conditions and settings), HPLC columns, buffers, and any other supplies needed for other labs to duplicate the validated method.

   b. Provide specific suppliers, addresses, catalog numbers, batch numbers, purity level, and any other unique identifying features that will ensure that other analysts obtain the exact items to duplicate the method(s).

   c. Indicate alternative commercial instruments and/or suppliers to be used in place of those specified by name.

   d. Provide any additional considerations necessary for sample preparation and workup, including factors such as dilution solvents, injection volumes, and gradient conditions. Specify the computer software and version used for data acquisition, manipulation, conversion, calculations, plotting, and comparison with library spectra.

   e. Indicate how all calculations, including statistical treatments, were performed.

f. Ensure that all necessary and sufficient details of the method are stated so that other labs can reproduce, as closely as possible, the experimental conditions.

13. *Transfer of Method Technology to Outside Laboratories: Interlaboratory Collaborative Studies*

a. Continue method validation (ruggedness) outside the original laboratory by performing interlaboratory collaborative studies. Interlaboratory studies can be accomplished by splitting known, authenticated samples and dispensing them to other laboratories while providing them with a complete procedure of the overall, final method.

b. Provide code numbers for actual samples and blanks (include blanks and placebos) so that data can be correlated for each sample. Indicate number of replicates desired and statistical treatment(s) to be employed.

c. Request that analysts in other labs report the total time required per sample for all samples analyzed, costs incurred, and any problems encountered.

14. *Comparison of Interlaboratory Collaborative Studies*

a. Summarize and statistically compare validation results from interlaboratory collaborative studies to demonstrate whether the method can be transferred to other facilities and provide similar accuracy and precision of the quantitative results.

b. Document ruggedness from lab to lab, using different analysts, experimental periods, instruments, chemicals, reagents, HPLC columns, and so on.

15. *Preparation of Summary Report on Overall Method Validation Results*

a. Prepare a summary report. Include results from all laboratories where the method was employed, with qualitative and quantitative results statistically treated. Summarize all the studies and final method performance.

b. It may be necessary to revise the original protocols and procedures established for the method based on the results of the interlaboratory ruggedness study. Detail any necessary changes, sources of reagents or instruments in addition to those first described, and any other comments derived from the collaborative study.

c. Provide each laboratory participating in the inter-laboratory collaborative study with the final report. This summary report allows any other suitably equipped laboratory to accurately and precisely perform the final, validated analytical method.

d. Obtain necessary management sign-off and observe any internal requirements to have the report accepted as a company's standard operating procedure (SOP).

16. *Summary Report of Final Method and Validation Procedures and Results, and Preparation of Journal Article for Submission*

a. Prepare a written summary of the final method and validation procedures, with optimized chromatograms and/or spectra that are suitable for publication in a recognized journal. These include the *Journal of the Association of Official Analytical Chemists, Pharmacopeia Forum, Analytical Chemistry, Journal of Chromatography,* and other related, peer-reviewed, methods-type journals.
b. Have the paper peer reviewed by other analysts not directly involved in any of the methods development, optimization, or validation steps.
c. Make any necessary changes in the method as a result of the peer review.

③

Method Validation (USP/ICH)

3.1 TERMINOLOGY AND DEFINITIONS

Method validation, according to the *United States Pharma-copeia* (USP), is performed to ensure that an analytical methodology is accurate, specific, reproducible, and rugged over the specified range that an analyte will be analyzed. Method validation provides an assurance of reliability during normal use and is sometime described as the process of providing documented evidence that the method does what it is intended to do. Regulated laboratories must perform method validation in order to be in compliance with FDA regulations. In a 1987 guideline (Guideline for Submitting Samples and Analytical Data for Methods Validation), the FDA designated the specifications in the current edition of the USP as those legally recognized when determining compliance with the Federal Food, Drug and Cosmetic Act. For method validation, these specifications are listed in USP Chapter <1225> and can be referred to as the "eight steps of method validation," as shown in Figure 3.

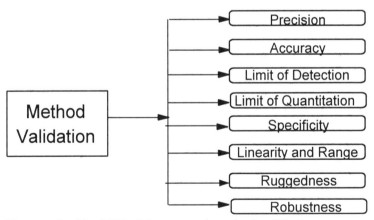

FIGURE 3 The USP eight steps of method validation.

These terms are referred to as "analytical performance parameters" or sometimes as "analytical figures of
merit." Most of these terms are familiar and are used
daily in the laboratory. However, some may mean different things to different people. To avoid confusion,
therefore, it is necessary to have a complete understanding of the terminology and definitions. Recognizing this,
one of the first harmonization projects taken up by the
ICH was the development of the guideline "Validation of
Analytical Methods: Definitions and Terminology." ICH
divided the validation characteristics somewhat differ-
ently, as outlined in Figure 4.

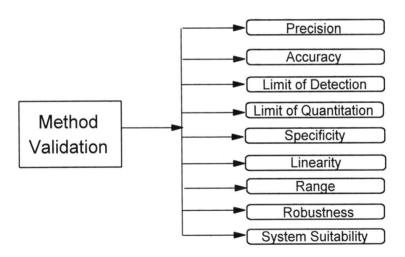

FIGURE 4 ICH method validation parameters.

The differences between the USP and ICH terminology is, for the most part, one of semantics—with one notable exception. ICH treats system suitability as a part of method validation, whereas the USP treats it in a separate chapter (<621>). Since this guideline has reached Step 5 of the ICH process, the FDA has begun to implement it, and it is anticipated that the ICH definitions and terminology will eventually be published in the USP. What follows then is a discussion of current USP definitions of the analytical performance parameters, compared and contrasted to the ICH definitions. Where appropriate, methodology is also presented according to the ICH methodology guideline.

3.1.1 Accuracy

Accuracy is the measure of exactness of an analytical method, or the closeness of agreement between the measured value and the value that is accepted either as a conventional, true value or an accepted reference value. Accuracy is measured as the percentage of analyte recovered by assay, by spiking samples in a blind study. For the assay of a drug *substance*, accuracy measurements are obtained by comparison of the results with those of a standard reference material, or by comparison to a second, well-characterized method. For the assay of a drug *product*, accuracy is evaluated by analyzing synthetic mixtures spiked with known quantities of components. For the quantitation of impurities, accuracy is determined by analyzing samples (drug substance or drug product) spiked with known amounts of impurities. [If impurities are not available, see Section 3.1.3 (p. 60) on specificity.)

To document accuracy, the ICH guideline on methodology recommends collecting data from a minimum of nine determinations over a minimum of three concentration levels covering the specified range (for example, three concentrations with three replicates each). The data should be reported as the percent recovery of the known, added amount, or as the difference between the mean and true value with confidence intervals. Accuracy can be documented through the use of control charts, an example of which is shown in Figure 5.

3.1.2 Precision

Precision is the measure of the degree of repeatability of an analytical method under normal operation and is normally expressed as the percent relative standard deviation for a statistically significant number of samples.

According to the ICH, precision should be performed at three different levels: repeatability, intermediate precision, and reproducibility. Repeatability refers to the results of the method operating over a short time interval under the same conditions (inter-assay precision). It should be determined from a minimum of nine determinations covering the specified range of the procedure (for example, three levels with three repetitions each), or from a minimum of six determinations at 100% of the test

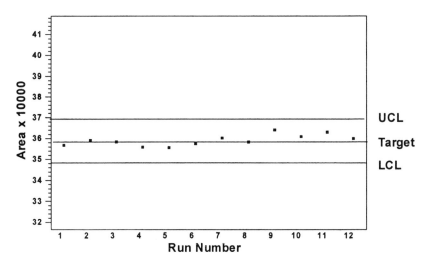

FIGURE 5 Documenting accuracy with Waters Millennium®
Chromatography Manager control charts.

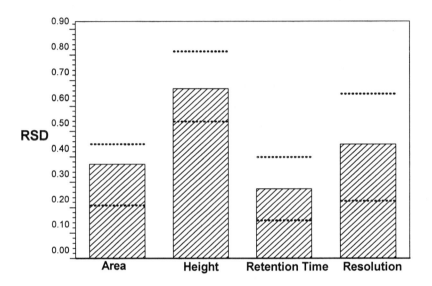

FIGURE 6 Documenting precision with Waters Millennium®
Chromatography Manager summary graphics bar plot custom
report.

or target concentration. Intermediate precision refers to the results from within-lab variations due to random events such as differences in experimental periods, analysts, equipment, and so forth. In determining intermediate precision, experimental design should be analyzed so that the effects (if any) of the individual variables can be monitored. Reproducibility refers to the results of collaborative studies among laboratories. Documentation in support of precision studies should include the standard deviation, relative standard deviation, coefficient of variation, and confidence interval. Figure 6 illustrates how custom graphics in the form of bar charts can be used to document precision.

3.1.3 Specificity

Specificity is the ability to measure accurately and specifically the analyte of interest in the presence of other components that may be expected to be present in the sample matrix. It is a measure of the degree of interference from such things as other active ingredients, excipients, impurities, and degradation products, ensuring that a peak response is due only to a single component; that is, that no co-elutions exist. Specificity is measured and documented in a separation by the resolution, plate count (efficiency), and tailing factor. Specificity can also be evaluated with modern photodiode array detectors that compare spectra collected across a peak mathematically as an indication of peak homogeneity.

ICH divides the term *specificity* into two separate categories: identification and assay/impurity tests. For identification purposes, specificity is demonstrated by the ability to discriminate between compounds of closely related structures, or by comparison to known reference materials. For assay and impurity tests, specificity is demonstrated by the resolution of the two closest eluting compounds. These compounds are usually the major component or active ingredient and an impurity. If impurities *are* available, it must be demonstrated that the assay is unaffected by the presence of spiked materials (impurities and/or excipients). If impurities are *not* available, the test results are compared to a second well-characterized procedure. For assay tests, the two results are compared; for impurity tests, the impurity profiles are compared head to head.

3.1.4 Limit of Detection

The limit of detection (LOD) is defined as the lowest concentration of an analyte in a sample that can be *detected*, though not necessarily quantitated. It is a limit test that specifies whether or not an analyte is above or below a certain value. It is expressed as a concentration at a specified signal-to-noise ratio, usually a 2- or 3-to-1 ratio. The ICH has recognized the signal-to-noise ratio convention but also lists two other options to determine LOD: visual noninstrumental methods and a means of calculation. Visual noninstrumental methods may include techniques such as thin-layer chromatography (TLC) or titrations. LODs may also be calculated based on the standard deviation (SD) of the response and the slope (S) of the calibration curve at levels approaching the LOD according to the formula: LOD = 3.3(SD/S). The standard deviation of the response can be determined based on the standard deviation of the blank, on the residual standard deviation of the regression line, or the standard deviation of y-intercepts of regression lines. The method used to determine LOD should be documented and supported, and an appropriate number of samples should be analyzed at the limit to validate the level.

3.1.5 Limit of Quantitation

The limit of quantitation (LOQ) is defined as the lowest concentration of an analyte in a sample that can be determined with acceptable precision and accuracy under the stated operational conditions of the method. Like LOD, LOQ is expressed as a concentration, with the precision and accuracy of the measurement also reported. Sometimes a signal-to-noise ratio of 10-to-1 is used to determine LOQ. This signal-to-noise ratio is a good rule of thumb, but it should be remembered that the determination of LOQ is a compromise between the concentration and the required precision and accuracy. That is, as the LOQ concentration level decreases, the precision decreases. If greater precision is required, a higher concentration must be reported for LOQ. This compromise is dictated by the analytical method and its intended use.

The ICH has recognized the 10-to-1 signal-to-noise ratio as typical, and as for LOD, lists the same two additional options that can be used to determine LOQ: visual noninstrumental methods and a means of calculation. The calculation method is again based on the standard deviation (SD) of the response and the slope (S) of the calibration curve according to the formula LOQ = 10(SD/S). Again, the standard deviation of the response can be determined based on the standard deviation of the blank, on the residual standard deviation of the regression line, or the standard deviation of y-intercepts of regression lines. As with LOD, the method used to determine LOQ should be documented and supported, and an appropriate number of samples should be analyzed at the limit to validate the level.

One additional detail should be considered: both the LOQ and the LOD can be affected by the chromatography. Figure 7 shows how efficiency and peak shape can affect the signal-to-noise ratio. Sharper peaks result in a higher signal-to-noise ratio, resulting in lower LOQs and LODs. Therefore, the chromatographic determination of LOQ and LOD should take into account both the type and age of the column, which is usually determined over the course of time as one gains experience with the method.

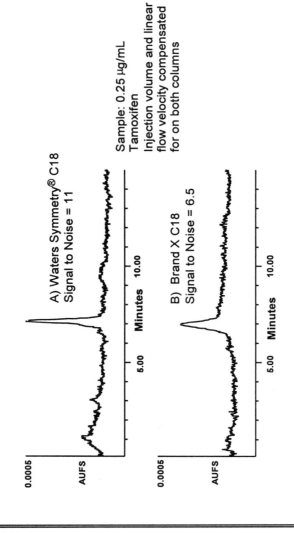

FIGURE 7 Effect of peak shape on LOD/LOQ.

3.1.6 Linearity and Range

Linearity is the ability of the method to elicit test results that are directly proportional to analyte concentration within a given range. Linearity is generally reported as the variance of the slope of the regression line. *Range* is the (inclusive) interval between the upper and lower levels of analyte that have been demonstrated to be determined with precision, accuracy, and linearity using the method. The range is normally expressed in the same units as the test results obtained by the method. The ICH guidelines specify a minimum of five concentration levels, along with certain minimum specified ranges. For assay tests, the minimum specified range is 80–120% of the target concentration. For impurity tests, the minimum range is from the reporting level of each impurity to 120% of the specification. (For toxic or potent impurities, the range should be commensurate with the controlled level.) For content uniformity testing, the minimum range is 70–130% of the test or target concentration, and for dissolution testing, ±20% over the specified range of the test. That is, in the case of an extended-release product dissolution test, with a Q-factor of 20% dissolved after six hours, and 80% dissolved after 24 hours, the range would be 0–100%. Figure 8 is an example of a linearity (calibration) plot.

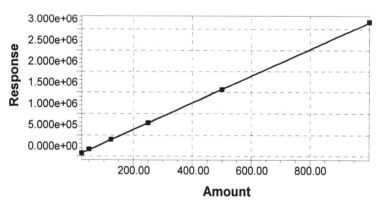

Processing Method : ANALGESIC_PM_15CM
System : MD2_LC_PDA   **Channel : 270nm**   **Date : 01-APR-94**
Type : LC   **Name : ACE**   **Retention Time : 8.402**
Order : 1   **A : -2310.580021**   **B : 3156.343178**
F :   0.000000   **R : 0.999979**   **R^2 : 0.999958**
Standard Error : 8519.681479

FIGURE 8   Waters Millennium® Chromatography Manager calibration plot.

3.1.7 Ruggedness

Ruggedness, according to the USP, is the degree of reproducibility of the results obtained under a variety of conditions, expressed as % relative standard deviation (RSD). These conditions include differences in laboratories, analysts, instruments, reagents, and experimental periods. In the guideline on definitions and terminology, the ICH does not address ruggedness specifically. This apparent omission is really a matter of semantics, however, as ICH chooses instead to cover the topic of ruggedness as part of precision, as discussed previously.

3.1.8 Robustness

Robustness is the capacity of a method to remain unaffected by small *deliberate* variations in method parameters. The robustness of a method is evaluated by varying method parameters such as percent organic solvent, pH, ionic strength, or temperature, and determining the effect (if any) on the results of the method. As documented in the ICH guidelines, robustness should be considered early in the development of a method. In addition, if the results of a method or other measurements are susceptible to variations in method parameters, these parameters should be adequately controlled and a precautionary statement included in the method documentation.

3.2 DATA ELEMENTS REQUIRED FOR ASSAY VALIDATION

Both the USP and the ICH recognize that it is not always necessary to evaluate every analytical performance parameter. The type of method and its intended use dictates which parameters need to be investigated, as illustrated in Table 1.

The USP divides analytical methods into three separate categories:

1. Quantitation of major components or active ingredients.
2. Determination of impurities or degradation products.
3. Determination of performance characteristics.

For assays in category 1, LOD and LOQ evaluations are not necessary because the major component or active ingredient to be measured is normally present at high levels. However, since quantitative information is desired, all of the remaining analytical performance parameters are pertinent. Assays in category 2 are divided into two subcategories: quantitative and limit tests. If quantitative information is desired, a determination of LOD is not necessary, but the remaining parameters are required. The situation is reversed for a limit test. Since quantitation is not required, it is sufficient to measure the LOD and demonstrate specificity and ruggedness.

TABLE 1   USP Data Elements Required for Assay Validation

Analytical Performance Parameter	Assay Category 1	Assay Category 2		Assay Category 3
		Quantitative	Limit Tests	
Accuracy	Yes	Yes	*	*
Precision	Yes	Yes	No	Yes
Specificity	Yes	Yes	Yes	*
LOD	No	No	Yes	*
LOQ	No	Yes	No	*
Linearity	Yes	Yes	No	*
Range	Yes	Yes	*	*
Ruggedness	Yes	Yes	Yes	Yes

*May be required, depending on the nature of the specific test.

TABLE 2 ICH Validation Characteristics Versus Type of Analytical Procedure

Type of Analytical Procedure	Identification	Impurity Testing		Assay
		Quantitative	Limit Tests	
Accuracy	No	Yes	No	Yes
Precision				
Repeatability	No	Yes	No	Yes
Interm. precision	No	Yes	No	Yes
Specificity	Yes	Yes	Yes	Yes
LOD	No	No	Yes	No
LOQ	No	Yes	No	No
Linearity	No	Yes	No	Yes
Range	No	Yes	No	Yes

The parameters that must be documented for methods in USP assay category 3 are dependent upon the nature of the test. Dissolution testing, for example, falls into this category.

The ICH treats analytical methods in much the same manner, as shown in Table 2. USP categories 1 and 2 match the ICH categories of assay and impurity testing, respectively, and the corresponding discussion above still applies. The ICH has not yet chosen to specifically address methods for performance characteristics (USP category 3), but instead has addressed analytical methods for compound identification. In this ICH category, it is necessary only to prove that the method is specific for the compound being identified.

④

System Suitability

According to the USP, system suitability tests are an integral part of chromatographic methods. These tests are used to verify that the resolution and reproducibility of the system are adequate for the analysis to be performed. System suitability tests are based on the concept that the equipment, electronics, analytical operations, and samples constitute an integral system that can be evaluated as a whole.

73

System suitability is the checking of a system to ensure system performance before or during the analysis of unknowns. Parameters such as plate count, tailing factors, resolution and reproducibility (%RSD retention time and area for repetitive injections) are determined and compared against the specifications set for the method. These parameters are measured during the analysis of a system suitability "sample" that is a mixture of main components and expected by-products.

Documentation of system suitability can be accomplished by using software specifically designed for the task to provide a review of the separation and to summarize the data regarding reproducibility. The software can also be used to troubleshoot the method. Results stored in a relational database can be compared and summarized on a peak-by-peak or system-by-system basis to provide the feedback necessary to determine system performance. Bar charts [see Figure 6 (p. 59)] can then be employed to document the required parameters.

Method Validation Protocol

USP Chapter <1225>, on validation of analytical methods, specifically addresses terms and definitions but leaves protocol and methodology open to interpretation. This omission may be intentional so as to allow flexibility in method validation. Laboratory personnel traditionally have been given latitude to do "good science"; however, the term "good science" means different things to different people. This is why the guidelines drafted as a result of the ICH process have gained importance. The ICH Guideline on Method Validation Methodology (now at Step 4 of the ICH process as of this writing) hints at experimental design and protocol. Before outlining an experimental design or protocol, however, it is necessary to make some basic assumptions:

1. Selectivity has been previously demonstrated, or is measured and documented during the course of the validation protocol.
2. The method has been developed and optimized to the point where it makes sense investing time and effort in validating the method. Indeed, robustness should be the first parameter investigated.
3. Once data is generated, statistically valid approaches are used to evaluate it and make decisions, thus lessening the subjectivity of method validation.

Given that the above three steps have already been addressed in one fashion or another, and bearing in mind the ICH methodology guidelines and that the requirements for validation depend on the type of analytical method [Tables 2 and 3 (pp. 69 and 70)], the following stepwise protocol can be proposed, as illustrated in Figure 9 (pp. 78–79):

1. On day 1, a linearity test over 5 levels for both the drug substance (bulk) and dosage form is performed.

2. Comparison of the results between the drug substance and dosage form fulfills the accuracy requirement.
3. At the end of day 1, 6 repetitions are performed at 100% of the drug substance for repeatability.
4. Steps 1 and 2 are repeated over 2 additional days for intermediate precision.
5. LOQ is evaluated (as needed) by analyzing the drug substance over 5 levels, plus 6 repetitions for precision.
6. Baseline noise is evaluated over 6 repetitions of blank injections for the determination of LOD (if required).

In this manner, using ICH methodology, a logical stepwise approach to method validation can be performed. The use of appropriate statistical tests (e.g., Student's *t*, Cochran, Dixon, and Fisher tests) allows for decisions regarding the data to be less subjective, making method validation a much more objective undertaking. A rigorous use of statistics may, in turn, lead to future automation of the decision process.

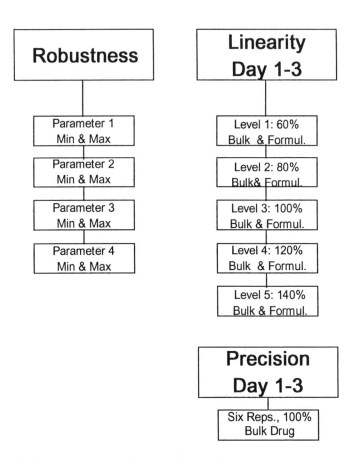

FIGURE 9 Stepwise protocol for method validation.

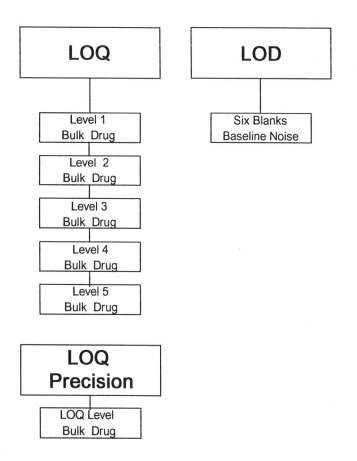

⑥

Method Transfer and Revalidation

After a method has been validated, it is ready to be transferred to other laboratories that can use the method. Ideally, laboratories are in constant communication so that all the goals of method development, optimization, and validation are achieved in an efficient manner. Once the agreed-upon goals have been accomplished, the method can be transferred to the end user.

Documentation of the method should include a detailed written procedure, a method validation report, system suitability criteria, and a plan for the method's implementation. General acceptance criteria for the method's intended purpose should also be included. The end user should plan on verifying method performance prior to making it an SOP. If the end user has been involved in the overall process of validation, either as an observer or as a participant (e.g., interlaboratory collaboration), the method can be introduced in a timely fashion.

At some time during the lifetime of a method, it may become necessary to revalidate the method. Revalidation can be carried out in a reactive or proactive manner. Reactive revalidation may be performed in response to changes in incoming raw material, manufacturing batch changes, formulation changes, or other changes (dilutions, sample preparation) in the method. A total revalidation from scratch is usually not called for in these instances, but enough revalidation to address the issues at hand, as dictated by the needs and use of the method, must be performed. Proactive revalidation may be undertaken to take advantage of new technology or to automate a labor-intensive and/or time-consuming manual procedure. In such cases, revalidation may be more comprehensive, depending on the undertaking.

Summary and Conclusions

Validation is a constant, evolving process that starts before an instrument is placed on-line and continues long after method development and transfer. A well-defined and documented validation process provides regulatory agencies with evidence that the system and method is suitable for its intended use. By approaching method development, optimization, and validation in a logical, stepwise fashion, laboratory resources can be used in a more efficient and productive manner.

Bibliography

Altria, K. D. Quantitative aspects of the application of capillary electrophoresis to the analysis of pharmaceuticals and drug related impurities. *J. Chromatogr.* *646*, 245 (1993).

Altria, K. D., Clayton, N. G., Hart, M., Harden, R. C., Hevizi, J., Makwana, J. V., and Portsmouth, M. J. An inter–company cross–validation exercise on capillary

85

electrophoresis testing of dose uniformity of para-cetamol content in formulations. *Chromatographia 39*, 180 (1994).

Bopp., R. J., Wozniak, T. J., Anliker, S. L., and Palmer, J. Development and validation of liquid chromatographic assays for the regulatory control of pharmaceuticals. In *Pharmaceutical and Biomedical Applications of Liquid Chromatography*, Riley, C. M., Lough, W. J., and Wainer, I. W., eds. Elsevier (Pergamon), 1994, pp. 315–344.

Buick, A. R., Doig, M. V., Jeal, S. C., Land, G. S., and McDowall, R. D. Method validation in the bioanalytical laboratory. *J. Pharm. Biomed. Anal. 8* (8–12), 629 (1990).

Carr, G. P., and Wahlichs, J. C. A practical approach to method validation in pharmaceutical analysis. *J. Pharm. Biomed. Anal. 8* (8–12), 613 (1990).

Debesis, E., Boehlert J. P., Givand T. E., and Sheridan J. C. Submitting HPLC methods to the compendia and regulatory agencies. *Pharm. Tech. 120*, (1982).

Development and Validation of Analytical Methods: Progress in Pharmaceutical and Biomedical Analysis, Volume 3. Riley, C. M., and Rosanske, T. W., eds. Elsevier (Pergamon), Tarrytown, NY, 1996.

Ding, X.-D., Magiera, D. J., Mazzeo, J. R., Edmars, K., and Krull, I. S. Determination of 2-acetyl-4(5)-(tetra-hydroxybutyl) imidazole in ammonia caramel color by HPLC. *J. Ag. Fd. Chem. 39*, 1954 (1991).

FDA's policy statement for the development of new stereoisomeric drugs. *Chirality 4*, 338 (1992).

Furman, W. B., Layloff, T. P., and Tetzlaff, R. F. Validation of computerized liquid chromatographic systems. *J. AOAC Int. 77* (5), 1314 (1994).

Green, J. M. A practical guide to analytical method validation. *Anal. Chem. News and Features*, May 1, pp. 305–309A (1996).

Guideline for submitting samples and analytical data for methods validation. Food and Drug Administration, February 1987. U.S. Government Printing Office: 1990-281-794:20818.

Henry, C. Reform and the well-characterized biologic. *anal. Chem. News and Features*, Nov. 1, p. 674A (1996).

Inman, E. L., Frischmann, J. K., Jimenez, P. J., Winkel, G. D., Persinger, M. L., and Rutherford, B. S. General method validation guidelines for pharmaceutical samples. *J. Chrom. Sci. 25*, 252 (1987).

International Conference on Harmonization. Draft guideline on validation of analytical procedures: Definitions and terminology. Federal Register, Vol. 60, p. 11260, March 1, 1995.

Karnes H. T., and March C. Calibration and validation of linearity in chromatographic biopharmaceutical analysis. *J. Pharm. Biomed. Anal. 9* (10–12), 911 (1991).

Karnes, H., Shiu, G., and Shah, V. P. Validation of bioanalytical methods. *Pharm. Res. 8* (4), 421 (1991).

Kirschbaum, J., Perlman, S., Joseph, J., and Adamovics, J. Ensuring accuracy of HPLC assays. *J. Chrom. Sci. 22,* 27 (1984).

Kirschbaum, J. J. Inter–laboratory transfer of HPLC methods: problems and solutions. *J. Pharm. Biomed. Anal. 7* (7), 813 (1989).

Krull, I. S., Mazzeo, J. R., and Selavka, C. M. *A rationale for analytical methodology development. Biomed. Chromatogr., 6,* 259 (1992).

Layloff, T., and Motise, P. Selection and validation of legal reference methods of analysis for pharmaceutical products in the U.S. *Pharmaceutical Technology,* September, pp. 122–132 (1992).

Maxwell, W., and Sweeney, J. Applying the validation timeline to HPLC system validation. *LC/GC 12* (9), 678–682 (1994).

McLaughlin, G. M., Nolan, J. A., Lindahl, J. L., Palmieri, R. H., Anderson, K. W., Morris S. C., Morrison, J. A., and Bronzert T. J. Pharmaceutical drug separations by HPCE: Practical guidelines. *J. Liquid Chromatogr. 15* (6/7), 961 (1992).

Paul, W. L. USP perspectives on analytical methods validation. *Pharmaceutical Technology,* March, pp. 130–141 (1991).

Peng, G. W., and Chiou, W. L. Analysis of drugs and other toxic substances in biological samples for pharmacokinetic studies. *J. Chromatogr. 531*, 3 (1990).

Proceedings from the Third International Conference on Harmonization, Yokohama, Japan, December, 1995.

Reviewer guidance, validation of chromatographic methods. Center for Drug Evaluation and Research, Food and Drug Administration, 1994.

Riley, C. M., Lough, W. J., and Wainer, I. W., eds. *Pharmaceutical and Biomedical Applications of Liquid Chromatograph.* Elsevier (Pergamon), 1994.

Schoenmakers, P. J., and Mulholland, M. An overview of contemporary method development in liquid chromatography. *Chromatographia, 25* (8), 737 (1988).

Shah, V. P., Midha, K. K., Dighe, S., McGilveray, I. J., Skelly, J. P., Yacobi, A., Layloff, T., Viswanathan, C. T., Cook, C. E., McDowall, R. D., Pittman, K. A., and Spector, S. Analytical methods validation: Bioavailability, bioequivalence, and pharmacokinetic studies. *J. Pharm. Sci. 81* (3), 309 (1992).

Shah, V. P., Midha, K. K., Dighe, S., McGilveray, I. J., Skelly, J. P., Yacobi, A., Layloff, T., Viswanathan, C. T., Cook, C. E., McDowall, R. D., Pittman, K. A., and Spector, S. Analytical methods validation: Bioavailability, bioequivalence and pharmacokinetic studies. *Pharm. Res. 9* (4), 588 (1992).

Snyder, L. R. In *Practical HPLC Method Development*, Snyder, L. R., and Glajch, J., and Kirkland, J. J., eds. Wiley-Interscience, New York, 1988.

Swartz, M. E. Method development and selectivity control for small molecule pharmaceutical separations by CE. *J. Liq. Chrom. 14*, 923 (1991).

Swartz, M. E., Mazzeo, J. R., Grover, E. R., and Brown, P. R. Validation of enantiomeric separations by micellar electrokinetic capillary chromatography using synthetic chiral surfactants. *J. Chrom. A. 735*, 303 (1996).

Taylor, J. K. Validation of analytical methods. *Anal. Chem. 65*, 600 (1983).

Thomas, B. R., and Ghodbane, S. Evaluation of a mixed micellar electrokinetic capillary electrophoresis method for validated pharmaceutical quality control. *J. Liquid Chromatogr. 16* (9/10), 1983 (1993).

U.S. Pharmacopeia 23, pp. 1982–1984, 1776–1777. U.S. Pharmacopeia Convention, 1995.

Vanderwielen, A. J., and Hardwidge, E. A. Guidelines for assay validation. *Pharm. Tech. 66* (March), (1982).

Virlichie, J. L., and Ayache, A. A ruggedness test and its application for HPLC validation. *S. T. P. Pharma Pratiques 5* (1), 49–60 (1995).

Watzig, H., and Dette, C. Precise quantitative capillary electrophoresis: Methodological and instrumental aspects. *J. Chromatogr. 636*, 31 (1993).

Weinberger, R. *Practical Capillary Electrophoresis.* Academic Press, San Diego, CA, 1993.

Williams, D. R.   An overview of test method validation. *BioPharm.* *34* (November), (1987).

Wilson, T. D. Liquid chromatographic methods validation for pharmaceutical products. *J. Pharm. Biomed. Anal. 8* (5), 389 (1990).